SCIENTIFIC AMERICAN EDUCATIONAL PUBLISHING

DIVE IN!

10 FUN EXPERIMENTS USING WATER

I0795214

BRING SCIENCE HOME

Published in 2024 by The Rosen Publishing Group, Inc.
2544 Clinton Street, Buffalo, NY 14224

Contains material from Scientific American , a division of Springer Nature America, Inc., reprinted by permission as well as original material from The Rosen Publishing Group.

Copyright © 2024 by The Rosen Publishing Group, Inc.

First Edition

All rights reserved. No part of this book may be reproduced in any form without permission in writing from the publisher, except by a reviewer.

Editor: Kristen Rajczak Nelson
Designer: Rachel Rising

All illustrations by Continuum Content Solutions

Activity on p. 5 by Science Buddies (December 17, 2015); p. 11 by Science Buddies, Svenja Lohner (January 9, 2020); p. 17 by Science Buddies (December 10, 2015); p. 22 by Science Buddies, Ben Finio (June 15, 2017); p. 26 by Science Buddies, Sabine De Brabandere (February 13, 2020); p. 31 by Science Buddies (November 5, 2015); p. 39 by Science Buddies (July 2, 2015); p. 45 by Science Buddies (April 10, 2014); p. 50 by Science Buddies, Svenja Lohner (July 5, 2018); p. 55 by Science Buddies (April 12, 2012).

Photo Credits: pp. 4, 5, 11, 17, 22, 26, 31,3 9, 45, 50, 55 Anna Frajtova/Shutterstock.com; pp. 5, 8, 9, 17, 20, 22, 24, 31, 36, 39, 43, 50, 53 cve iv/Shutterstock.com.

Cataloging-in-Publication Data
Names: Scientific American, Inc.
Title: Dive in!: 10 fun experiments using water / edited by the Scientific American Editors.
Description: New York : Scientific American Educational Publishing, an imprint of Rosen Publishing, 2024. | Series: Bring science home | Includes glossary and index.
Identifiers: ISBN 9781725349919 (pbk.) | ISBN 9781725349926 (library bound) | ISBN 9781725349933 (ebook)
Subjects: LCSH: Hydrology--Experiments--Juvenile literature. | Water--Experiments--Juvenile literature.
Classification: LCC GB658.35 D58 2024 | DDC 333.91'4--dc23

Some of the images in this book illustrate individuals who are models. The depictions do not imply actual situations or events.

CPSIA Compliance Information: Batch #CWSA24. For further information contact Rosen Publishing, New York, New York at 1-800-237-9932.

Find us on

CONTENTS

⚛ THESE ACTIVITIES INCLUDE SCIENCE FAIR PROJECT IDEAS.

INTRODUCTION

The vast majority of Earth—71 percent—is made up of water. So is the human body—between 55 percent and 60 percent! Water is integral to life on this planet and is the source and subject of many scientific discoveries. The following experiments will show you how water affects the world all around us!

Projects marked with ⚛ include a section called Science Fair Project Ideas. These ideas can help you develop your own original science fair project. Science fair judges tend to reward creative thought and imagination, and it helps if you are really interested in your project. You will also need to follow the scientific method. See page 61 for more information about that.

Capable Carbon Filters

CAN YOU CLEAR YOUR COLORED DRINKS? LEARN HOW A SPECIAL TYPE OF CARBON CAN CLEAN OUT IMPURITIES FROM YOUR DRINKING WATER—AND WATCH AS IT SUCKS UP THE DYE FROM A SPORTS DRINK RIGHT BEFORE YOUR EYES!

PROJECT TIME

30 to 45 minutes

KEY CONCEPTS

Adsorption
Activated carbon
Physics
Chemistry
Water filter

Do you filter your tap water before drinking it? Maybe at home you have a water filter on your faucet, in the fridge, or use special pitchers that have a filter in them. You have probably heard commercials that claim these filters make your drinking water cleaner and safer. But have you ever wondered what, exactly, these filters do and if the water is really cleaner in the end? Find out about the cleaning power of water filters in this activity—but with results you can see. You will "clean" a colored sports drink or colored soda!

BACKGROUND

To find out how water filters work it is probably best to have a look inside the filter. Most of the filters that are used for home water treatment are carbon filters. That means the material inside the filter is carbon or a special form of it, called activated carbon. What makes activated carbon special is that it is a very porous form of carbon—almost like a sponge—and has many tiny, microscopic pores that soak up water. All these little micropores create a huge inner-surface area. About 5 teaspoons of granular activated carbon has a surface area that is approximately the area of a football field.

When water or liquid travels through the porous structure of the filter, impurities (such as small amounts of chemicals or metals) can be removed by a process called adsorption. Adsorption occurs when compounds physically or chemically adhere to the carbon surface and a film of the adsorbate, or the chemical impurities, is created on the adsorbent (carbon surface). This is also why the surface area matters. The more surface area, the more possible bonding sites there are for the impurities. If all the bonding sites are taken up, then the impurities remain in the water, and it is time to replace your water filter. Besides the surface area, the time the water spends in contact with the activated carbon is also an important factor that determines the efficiency of the filtration process. The longer the contact time or the slower the flow rate of the water, the more adsorption can take place. See for yourself in this activity. It's time to get started with your own adsorption experiment!

MATERIALS

- Colored sports drink or colored soda
- Five plastic cups (2-ounce [59 ml] cups work best)
- Tablespoon
- Activated carbon (You can get this from a pet store. If you would like to taste your results, use food-grade activated carbon.)
- Timer
- Coffee filters (at least four)
- Permanent marker
- Paper towels
- White sheet of paper
- A working area that can tolerate some liquid splashes

PREPARATION

- Label five 2-ounce (59 ml) plastic cups with: 0, 1a, 1b, 2a, and 2b.
- Using 1/2 tablespoon of activated carbon, fill cup 1a. Do the same for 2a.
- Prepare two double-layered coffee filters. Insert one filter into another filter to form the two layers.
- Get your timer and colored sports drink ready.

PROCEDURE

- Take the empty 2-ounce (59 ml) plastic cup, labeled 0, and add about 1 tablespoon (15 ml) of colored sports drink or soda. *How does the drink look inside the cup? Is the color very intense? Smell the sports drink in the cup. Does it have a specific smell?*
- Set your timer for five minutes. Do not start it yet.
- To the two cups with activated carbon, labeled 1a and 2a, add about 1/2 tablespoon (7 ml) of colored sports drink to each cup. Carefully swirl each cup to make sure that the solution mixes well with the activated carbon. *What happens when you add the solution to the activated carbon? What does the mixture look like?*
- Start your timer immediately after you have filled both cups.
- Take the double-layered coffee filter and hold it above cup 1b. Once the timer rings after five minutes, pour the contents of cup 1a (liquid and activated carbon) into the double-layered coffee filter. Collect the liquid that runs through the filter in cup 1b. *Did the appearance of the sports drink change? How?* Smell the liquid again. *Do you notice a difference compared with the original sports drink?*
- Set your timer for 25 minutes and start it immediately after you finish filtering the solution from cup 1a.

- While waiting, swirl cup 2a occasionally.
- Now prepare your second double-layered coffee filter, and once your timer rings after 25 minutes, hold it over cup 2b. Pour the liquid and activated carbon from cup 2a into the filter and collect the liquid in cup 2b. *How does this solution look compared with the original sports drink and the solution in cup 1b? Do you notice any change in smell?*
- Finally, line up cup 0 with your original sports drink, cup 1b, and 2b. Put them on a white sheet of paper so you can see the solutions better. Look at all three cups. *How did the sports drink change over time when it was in contact with the activated carbon?*

EXTRA

If (and only if) you used food-grade activated carbon, you can taste each of your solutions. *Do you notice any differences compared with the original sports drink? How does the taste change?*

SCIENCE FAIR IDEA

Try different colors of sports drinks or sodas. *Do you see the same effect for all of them or does one drink work better than the others?*

SCIENCE FAIR IDEA

What happens if you change the ratio of activated carbon and liquid? Try this activity with different amounts of activated carbon in the same volume of sports drink. Or change the volume of your solution and keep the amount of activated carbon the same. *Do different activated carbon-liquid ratios lead to different results? What do you see after five minutes, for example, when you add more sports drink to the same amount of activated carbon?*

SCIENCE FAIR IDEA

Try to find other materials instead of activated carbon that could clean your solution. *Can you think of other adsorbing materials that would work the same way as activated carbon? What property does such a material need to have?*

OBSERVATIONS AND RESULTS

Sports drinks come in a variety of colors. Looking at your plain drink in the plastic cup, you probably found that the color was very intense. Depending on the color or flavor, you might have noticed a fruity or sweet smell. When you poured your drink into the cups with activated carbon, did you see it fizzing and bubbling? This is because once the activated carbon becomes wet, all the air that is inside its many pores is replaced by liquid and, therefore, is released, which leads to the bubbling and fizzing.

The mixture now looks like a black slurry, and if you filtered it into cup 1b after five minutes, you probably noticed that the original color of the solution had become much less intense. And the fruity smell probably was very faint or not even recognizable anymore. If you check the ingredients list of your sports drink or soda, you probably will find that there are specific ingredients listed that make the color and flavor, such as "Blue no. 1," "caramel color," or "artificial flavor." These are mostly chemicals or organic compounds that will adsorb to the activated carbon surface and get trapped in its pore spaces. Thus, they will be removed from your drink. When treating your sports drink longer with the activated carbon, you can even remove all its color so you end up with an almost clear solution. Did your sports drink look clear in cup 2b once you filtered out the activated carbon after 25 minutes? The longer the beverage is in contact with the activated carbon, the more its color and fragrance chemicals can be adsorbed to the carbon's surface. If you wait long enough, all the color and smell is gone.

If you play around with the ratio of liquid volume and amount of activated carbon, however, you might find that if you have a lot of liquid and only very little activated carbon, you will not be able to remove the color completely. This is because the activated carbon has a limited adsorption capacity, which means that once the surface area is completely covered with color and flavor compounds, no space remains to adsorb more. This is also the reason why you have to replace your water filters at home regularly.

If you used food-grade activated carbon and tasted the colorless sports drink, it probably did not taste like the original one at all! Now think of the water filter you use at home—it works exactly the same way. Many of the compounds (even some that you cannot see or taste) will be trapped in the activated carbon filter, and the only thing left is water for you to drink.

CLEANUP

You can pour all the solutions into the sink. Collect the activated carbon in one container and discard it with your regular trash. Wipe your work area down with wet paper towels.

Make a Potato Shrink —with Saltwater

CREATE MOVEMENT WITH SALT! LEARN HOW PLANT CELLS REGULATE WATER WITH AN ACTIVITY YOU CAN SEE—AND FEEL.

Have you ever wondered how plants "drink" water from the soil? Water uptake in plants is quite complicated. A process called osmosis helps the water move from the soil into the plant roots—and then into the plant's cells. In this activity, you will see for yourself how you can make water move with osmosis!

PROJECT TIME

45 to 60 minutes

KEY CONCEPTS

Biology
Osmosis
Cells
Chemistry
Concentration
Water transport

BACKGROUND

Most water in the ground is not pure water. It usually contains dissolved mineral salts. Animals and plants need these salts (which include calcium, magnesium, potassium, and the sodium you might be familiar with as table salt) to grow, develop, and stay healthy. Different water sources carry different amounts of these salts. Nature wants to balance a system that is not balanced. So if you mix water with two different salt concentrations, the salts don't stay separated but spread out evenly through the solution until the salt concentration is the same throughout.

You'll find a similar reaction if you separate two salt solutions with a semipermeable membrane. A semipermeable membrane is a type of barrier that only lets certain particles pass through while blocking others. This type of membrane usually lets water pass through but not the salts that are dissolved in the water. In this situation, because only water can move through this membrane, the water will start moving from the area of lower salt concentration (which has more water and less salt) to the area of higher salt concentration (which has less water and more salt). This water movement will only stop once the salt and water concentration on both sides of the membrane is the same.

The process of moving water across a semipermeable membrane is called osmosis. Plants use this process to their advantage for water uptake. They create an environment of high salt concentration in their root cells that are in contact with the soil. The cell walls act as a semipermeable membrane that only lets water through. Because the water outside the root cells has a lower salt concentration, water starts moving into the root cells due to osmosis. The water entering the plant fills up the cells and can travel to the rest of the plant. Osmosis, however, works in both directions. If you put a plant into water with a salt concentration that is higher than the concentration inside its cells, water will move out of the plant to balance out the concentration difference. As a result, the plant shrinks and eventually dies. You will see this effect with your own eyes in this activity using potatoes and different saltwater solutions.

MATERIALS

- Distilled water
- Measuring cup with milliliters (ml)
- Table salt
- Weight scale with gram measurements

- Three plastic cups or glasses
- Spoon
- At least three potatoes
- Apple corer (Alternatively, you can have an adult help you use a cutting board and knife.)
- Knife (and an adult helper to help you use it)
- Ruler with millimeter measurements
- Paper
- Pen or pencil
- Timer
- Paper towels
- Graphing paper (optional)
- Other vegetable(s) or fruit (optional)

PREPARATION

- Prepare three different saltwater solutions. Create labels for the three cups: "0 grams," "2 grams," and "4 grams."

- To each of the cups add 100 milliliters of distilled water.

- Weigh out 2 grams of table salt, and add it to the cup that says "2 grams." Then weigh out 4 grams of table salt, and add it to the cup labeled "4 grams." Use a spoon to mix the solutions until all the salt is dissolved.

- Draw a table in which you can put the starting measurements (length and diameter or width) and end measurements of each potato strip for every salt concentration (0, 2, and 4 grams).

- Prepare at least three potato cores. Carefully push the corer all the way through the potato, and remove the core carefully so the potato piece stays intact. (Alternatively, you can have an adult help cut the potato into strips that all have the same dimensions.) The potato pieces should be at least 12 millimeters thick and 50 millimeters long. (Ideally you will be able to prepare nine matching cores or strips so you can test three pieces in each solution to compare the results thoroughly.)

- Use a knife to carefully remove any potato skin from your cores, and rinse the cores quickly with water.

- Use a ruler to ensure each potato piece is the same size (ideally to the millimeter). Carefully use a knife to trim any pieces as needed.
- Measure the dimensions (length and diameter or width) of each potato strip in millimeters, and write the information in the table.
- Optionally, you can also weigh each potato piece and record their weights.

PROCEDURE

- Put one potato strip (or three if you made nine pieces) into each of the cups. While you do that, feel the potato strips with your fingers and try to flex them a little bit. *How do they feel? Are they easy to bend?*
- Start your timer for 30 minutes. Let the potato strips sit in the different solutions for the whole time. *What do you think will happen to the strips in each of the cups?*
- After 30 minutes, inspect the potato strips inside the solutions. *Do you see any changes?*
- Take the potato strip(s) out of the "0 grams" cup and place on a paper towel. While doing that, feel the potato pieces again and try to bend them slightly. *How do they feel? Are they easier or more difficult to bend than before?*
- Use the ruler to measure the exact length and diameter or width (in millimeters) of each of the potato strips, and write the results in your table. *What do you notice about the potato strip measurements?* Optionally you can weigh these pieces and record their weights.
- Next, take the potato strips from the "2 grams" cup, and place them on a paper towel. As you do this, feel them. Measure their lengths and diameters or widths. Write your results in the table. Optionally you can weigh these pieces and record their weights. *What changed about these potato strips?*

- Repeat the same steps with the potato strips in the "4 grams" cup. Write your results in the table. *Are your results for these similar or different compared with the other ones?*
- *How did the feeling of the strips compare based on what solution they were in? Why do you think this is?*
- Compare the results in your table. *How did the length and diameter or width of the potato strips change in each cup? What about the weights if you took them? Can you explain your results?*

EXTRA

If you weighed each of your strips before and after soaking them, compare the weights. *How does the mass of the potato strips change in each solution?*

EXTRA

Leave the potato strips in the solutions for a longer time period. *How do they look if you let them soak in the saltwater for one hour or overnight?*

EXTRA

If you have graphing paper, make a graph of your results with the salt concentration on the horizontal axis and the potato strip length or diameter after soaking on the vertical axis. Draw two lines to make your graph. For the first, connect each of the data points you found. For the second, draw a horizontal line starting at the point on the vertical axis that shows the original length of your potato strip. *Based on your graph, can you find a salt concentration at which the potato strip length should not change at all?*

EXTRA

How does the activity work with other vegetables or fruit? Try it to find out!

OBSERVATIONS AND RESULTS

Did your potato strips shrink and expand? At the beginning, all the potato strips should have had the same length and should have all felt the same. When you put them into the different solutions, however, this starts to change. Whereas the potato strips in the "0 gram" cup probably got larger in size, the other potato strips probably got shorter after leaving them in the saltwater for 30 minutes. (If you didn't see any significant changes after 30 minutes, leave the potato strips in the saltwater solutions longer.)

The shrinking and expanding of the potato strips is due to osmosis. Potatoes are made of cells, and their cell walls act as semipermeable membranes. The 0 gram solution contains less salts and more water than the potato cells (which have more salts and less water). To balance out these concentration differences, the water from the cup moves into the potato cells. The incoming water in the potato cells pushes on the cell walls and makes the cells bigger. As a result, the whole potato strip gets bigger. The opposite is the case in the higher concentration salt solutions. If the salt concentration in the cup is higher than inside the potato cells, water moves out of the potato into the cup. This leads to shrinkage of the potato cells, which explains why the potato strips get smaller in length and diameter. Due to the shrinking of the potato cells, the potato strip also becomes less rigid. If you bent the potato strips, you should have noticed that those that had been in the solution with the highest amount of salt were much easier to bend than the potato strips in the water without salt.

If you made the graph, you probably noticed that there is a salt concentration at which the potato strip neither expands nor shrinks. This should be where your data curve and your start length line intersect. At this point the salt concentration inside the potato cells and inside the cup are the same. Because the concentrations are already balanced, no water moves.

CLEANUP

Discard the saltwater solutions in the sink. Throw the potato strips into the compost, and clean up your workspace. You can cook with the other pieces of unused potato.

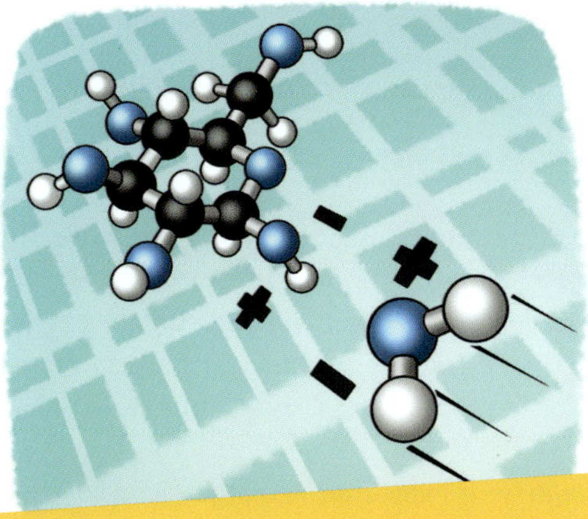

Folded or Flat Paper Towel
Which One Absorbs More Water?

A LITTLE SCIENCE CAN HELP YOU SAVE AN IMPORTANT RESOURCE YOU PROBABLY USE EVERY DAY: PAPER TOWELS. TRY THIS ACTIVITY AND SEE WHAT SIMPLE TRICK CAN HELP YOU—AND THE ENVIRONMENT!

PROJECT TIME

20 to 30 minutes

KEY CONCEPTS

Absorption
Paper
Water
Molecules

We all know that washing hands throughout the day can help keep colds and flu at bay. So several times a day we lather up our hands, scrub, rinse and then use a paper towel—then another one, maybe even three or four—to dry them off. Because who wants wet hands? But could there be a way to conserve some of that paper by getting a paper towel to go the extra mile, allowing you to dry your hands with just one single sheet? This activity just might help you find the answer.

BACKGROUND

To understand how paper towels absorb water, we need to know a little about how they are made.

Paper towels are made of ground-up plant material. If you look through a microscope at a torn-up piece of paper (or look up some images on the internet), you will see a web of tiny plant fibers. Magnifying your paper further will reveal that the fibers are made of long chains of linked sugar molecules, called cellulose. Water is attracted to cellulose and likes to be soaked up and stick to the cellulose in paper.

When you looked through your microscope, did you also see the spaces between the fibers? These empty spaces affect the absorbency of the paper: Water likes to stick together and fill these spaces as it follows the water attracted to the cellulose. More spaces allow more water to be absorbed. But what would happen if you add a tiny space between sheets of paper towels? Would the empty space between the sheets help to hold more water?

MATERIALS

- Five or more identical paper towels—preferably the type in public restrooms (In case you would like to test different types of paper towels, choose at least five identical towels of each type. Do not worry about wasting a few towels—this activity might help save paper in the long run!)
- Water
- Place to hang a paper towel to drip
- Kitchen scale, 1-gram precision or better
- Paper and pen or pencil (for recording weights)
- A workspace that can get a little wet
- Scissors to make towels smaller for small hands (optional)

PREPARATION

- Assemble all of your materials at your workspace.

PROCEDURE

- Unfold the first paper towel (if you have the prefolded type). Wet it thoroughly and hang it so all of the excess water drips out.
- When the towel no longer drips, weigh it on a kitchen scale. You can heap up the towel on the scale rather than neatly folding it. Record the mass on a piece of paper.
- Fold an identical paper towel in three (if it was not already prefolded) and fold it one more time so six layers of towel are on top of one another. Wet it thoroughly and hang it—still folded—so all of the excess water drips out. *Do you think this folded paper towel holds more, less, or just as much water as the unfolded paper towel?*
- When the folded towel stops dripping, weigh it on a kitchen scale. Do not unfold it; place it on the scale then read and record its mass. *Does it weigh more, less, or exactly the same as the wet unfolded paper towel? If there is a difference, why do you think the mass is different?*
- *Now that you measured how much water the folded and unfolded paper towels can hold, and maybe found a difference, which do you think would dry your hands better?*
- Place a fresh, unfolded paper towel and an identical fresh paper towel folded in three in a dry spot on your workspace.
- Wet your hands, shake them three times to remove most of the water, and then dry them off with the unfolded paper towel. *Do your hands feel completely dry, somewhat dry, or still quite wet?*
- Repeat wetting and shaking your hands. Try to shake your hands in the same way you did the first time, then dry them with the folded paper towel. *How do your hands feel now? Do they feel dryer, wetter, or just as dry as when you used the unfolded paper towel?*

- If your hands feel very dry with both the folded and unfolded paper towels, try again with half a paper towel, as follows: Cut a paper towel in half and dry your hands with an unfolded half-towel and with a folded half-towel. *Do you feel a difference now?*
- *How can your findings help you use fewer paper towels for the same job?*

EXTRA

If you have more paper towels of the same type, repeat the tests; perform each step exactly the same way and notice the variations in the outcomes. Does the measured difference in mass vary a lot or just a bit? *Is it always the folded or always the unfolded paper towel that weighs more? Do your hands always feel drier when using the folded or the unfolded paper towel?* Scientists repeat tests to verify the outcome. Scientists also like to have their studies repeated by a different researcher utilizing different instrumentation (such as the scale). If the independent tests reveal the same results, the test is called reproducible. Such repeated tests by other experimenters have more scientific value. *Can you find a friend to help you make your tests reproducible?*

SCIENCE FAIR IDEA

If you have different types of paper towels available, repeat the tests with them. *Do you expect similar results? What did you find after testing them all?*

SCIENCE FAIR IDEA

Test other paper products that are used to absorb liquids, such as kitchen paper towels, toilet paper, paper napkins, or tissues. *Do these absorb more water when folded than when used single-layered? Which type of paper product gains most by folding? Can you explain why?*

OBSERVATIONS AND RESULTS

Did you measure a higher mass for the folded wet paper towel and did your hands feel drier when you used it? This is expected, as the tiny space between paper towel layers helps hold more water.

Paper is made of cellulose, which water molecules like to cling to. As a result, paper readily absorbs water. Paper towels are especially absorbent because their cellulose fibers have empty spaces—tiny air bubbles—between them. Water molecules, which like to stay together, follow one another as they are absorbed by the cellulose and fill the empty spaces. Layering the paper towel creates more spaces for water to fill, which explains why your layered paper towel could hold more water and was more efficient at drying your hands.

The next time you reach for the paper towels, remember to fold! You might feel good knowing you just saved an extra paper towel from being used.

CLEANUP

Let your wet paper towels dry, then compost them if possible.

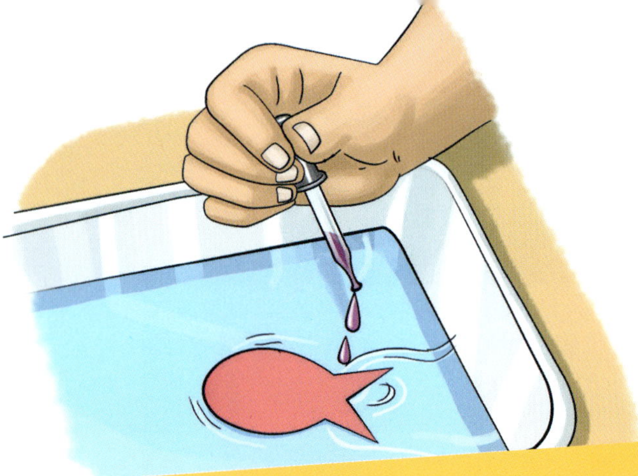

Make a Paper Fish Swim with Surface Tension

CAN YOU MAKE A TOY FISH THAT SWIMS ALONG WITH PHYSICS? TRY THIS HIGH-TENSION ACTIVITY AND FIND OUT!

Have you ever wondered how water droplets form on a window on a rainy day, why some bugs can walk on water, or why you use soap to wash your hands? All of these things are related to the phenomenon called surface tension. In this activity, you will learn how you can use surface tension to make a fish cut from a piece of paper seem to swim around on its own!

PROJECT TIME

20 to 30 minutes

KEY CONCEPTS

Physics
Forces
Molecules
Surface tension

BACKGROUND

Whether riding in a car on a rainy day or spilling water on the kitchen counter, you have probably seen water droplets form. What holds these water droplets together—sometimes in nearly spherical shapes? Why don't they spread out evenly or fall apart into even smaller drops? They are held together by something called surface tension. Water molecules are slightly attracted to one another; at the surface of a water droplet, they are more attracted to the other molecules around them than to the air above. This forms a thin "skin" of strongly attracted water molecules that are difficult to break through.

Sometimes high surface tension can be beneficial—such as for insects that walk on water. Other times we want to lower water's surface tension—for instance, when we are doing laundry or washing dishes. This is because there are lots of tiny holes in between pieces of dirt or clothing fabric, and high surface tension can make it hard for water to seep into those holes. Adding soap or detergent decreases water's surface tension, making it easier to penetrate all those little nooks and crannies. Do you think you can make use of this change in surface tension to make an object skim across the surface of the water? Try this project to find out!

MATERIALS

- Shallow tray
- Tap water
- Piece of paper
- Scissors
- Medicine dropper
- Soap
- Workspace that can tolerate soapy water spills

PREPARATION

- Cut out a small "fish" shape from your piece of paper. Make sure the fish has a V-shaped notch in its tail.
- Fill your shallow tray partially with water, and bring it to your workspace.

PROCEDURE

- Carefully place the paper fish on the surface of the water at one end of the tray with its tail facing outward and its head facing into the tray. Make sure the fish floats and you do not push it underwater.
- Use the medicine dropper to carefully place a single drop of soap on the water in the V-shaped notch of the fish's tail. *What happens?*
- Empty, rinse and dry the tray. Cut out a new paper fish, and try again. *Do you get the same result?*
- Try reusing one of the paper fish with fresh water, reusing the tray of water with a new paper fish or both. *Can you get it to work a second time? Why or why not?*

SCIENCE FAIR IDEA

Try cutting out and testing some different shapes. *Do the shapes have to be "streamlined" like a fish? What if the shape is symmetric like a circle or blocky like a square?*

SCIENCE FAIR IDEA

Try the activity with different liquids you have in your kitchen, such as vegetable oil or juice. *Do any liquids have the same effect as soap? Do any of them not work? Why do you think this happens?*

OBSERVATIONS AND RESULTS

When you place the paper fish on the water's surface and it floats, the surface tension of the water pulls at it from all sides. Because it is completely surrounded by water, surface tension pulls evenly in every direction, and it does not move. When you add a drop of soap to one side of the fish, the surface tension on that side suddenly becomes lower so the fish is not pulled as hard in that direction, and it moves in the other direction. This is a great example of Newton's first law of motion! The fish is initially at rest and the forces on it are balanced so it remains at rest. When you add the soap, the forces on it become unbalanced so it accelerates.

You will probably find it is difficult to either reuse a paper fish or a tray of water. Once the fish and/or water are "contaminated" with some soap, the soap spreads and the surface tension is lowered everywhere, rather than on just one side of the fish, so you can no longer make it move. You can probably find, however, the effect does work with many other shapes—even squares! You might also observe that other liquids, such as oils (which have a lower surface tension than water) work just as well as soap, but liquids such as juices (which are mostly water) do not.

CLEANUP

Clean up any spilled liquids.

Holes That Do Not Leak!

THERE'S NOT A DRIP TO SEE! LEARN HOW THE AMAZING POWERS OF PRESSURE CAN KEEP A WATER BOTTLE FROM LEAKING.

Did you know that at sea level there are about 15 pounds of air pressing on each square inch of your body? This air is very helpful in our daily lives. For example, the layer of air around Earth helps to keep it from getting too cold or hot. It can even help keep a bottle with holes in it leak-free! Try the activity to find out how.

PROJECT TIME

30 to 45 minutes

KEY CONCEPTS

Physics
Air pressure
Water pressure
Surface tension

BACKGROUND

Earth is covered by the atmosphere, which is a blanket of gas that is 60 miles (96.6 km) thick. Although we usually think of air as not being full of anything, all air is made up of tiny particles, which have a small amount of mass. We—and anything else around us—experience the weight of this layer of gas as pressure; this is called atmospheric pressure. We are so used to this pressure, however, that we rarely notice its existence. But if you have ever felt your ears "pop" while driving up a mountain, you've noticed it changing. As you drive up in elevation, fewer layers of air press on you. The air inside your eardrums remained at the air pressure from a lower elevation—at least, until they "popped."

Water is also made up of tiny particles that have mass. When something is underwater it feels the pressure of all the layers of water above it. Because water is much denser than air, it is also much heavier. A layer of about 33 feet (10 m) of water creates approximately the same pressure as the layer of air surrounding Earth. You might have felt water pressure while diving in a deep pool; the deeper you dive, the more you feel the water press against your eardrums.

Water has another interesting quality: its particles like to stay together. It is as if there is a thin film around a body of water. Scientists call this surface tension. You can see surface tension at work when looking closely at water droplets on a solid surface; they tend to clump together in round dots or small puddles rather than spreading out completely flat and evenly.

In this activity, you will find out how you can use air pressure and surface tension to keep a bottle with holes in it from leaking.

MATERIALS

- Sturdy plastic bottle with tight-fitting lid (25 ounces [740 ml] to 68 ounces [2 L] works well)
- Water (enough to fill the bottle)
- At least four pushpins
- Work area that can get wet
- Baking pan with a rim
- Sink
- Towel or cloth for cleanup
- More sturdy plastic bottles (optional)
- More water (optional)

PREPARATION

- Fill the bottle with water, and close the lid tightly.
- Place the bottle on the baking pan. The pan will catch any water that might flow out.

PROCEDURE

- Push at least four pushpins into the body of the bottle, about 1 inch (2.5 cm) from its bottom. *Does water leak out? Why do you think this is so? What do you think will happen if you pull the pins out?*
- Carefully pull the first pin straight out so you leave a small round hole in the bottle. *Do you see a stream of water flowing out or just a trickle? Why do you think that is so? What do you think will happen when you pull the other pins out?*
- Carefully pull the other pins out of the bottle. Try not to press on the bottle while you pull the pins out. *What happens? Is this what you expected? Why do you think this happens?*
- Press on the bottle and release it. Repeat this a few times. *What happens when you press on the bottle? What happens when you release it? Why do you think this would happen?*
- Carefully move the bottle to a sink. Hold a towel underneath to catch any drips. Hold the bottle over the sink, and open it. *What happens? Why would this happen? Can you find a way to stop the streams of water?*

EXTRA

In another bottle, make holes at different heights in the body of the bottle. All of the holes should be well below the waterline. *Do you think that making holes at different heights will alter the results?*

EXTRA

Try again but make your last hole near the bottle's spout so air can flow in and out of the bottle. *How does this small hole change the outcome? Why do you think that is?*

EXTRA

This activity asks to make small round holes in the bottle. Try repeating the activity with larger holes or irregularly shaped holes, and see if the results change. *Why would the size or shape of the hole matter?*

OBSERVATIONS AND RESULTS

It is likely that only a trickle dripped from the holes when the bottle remained closed and no pressure was applied—and that streams poured out when the bottle was opened or pushed in.

Water likes to stick together, so it takes effort to separate a stream of water from a body of water. Unless you press the bottle—or unless the air in the atmosphere pushes on the top surface of the water—no streams will flow out from tiny holes.

Air outside the bottle also presses against the water near the holes. If there is a place for air to flow in (such as an opening at the top of the bottle), the entering air could allow water to drip out of the holes. But in an airtight bottle, no air can enter to fill the space of any water that dripped out. So if any bit of water leaves the bottle, it reduces the air pressure inside the bottle, keeping the water from pouring out. If you squeeze the bottle, however, you increase the air pressure, and water can squirt out.

If the holes are small enough, the water sticks together just enough to stop air from bubbling in. If you tried making larger holes, you probably noticed that air bubbles manage to creep in as water flowed out of them. If you made a small hole near the spout of the bottle, you undoubtedly noticed that even a tiny hole can allow enough airflow in to get the water flowing.

CLEANUP

Pour out the water in a sink, and wipe up any water that's spilled. Recycle the plastic bottle, and return other items where you found them.

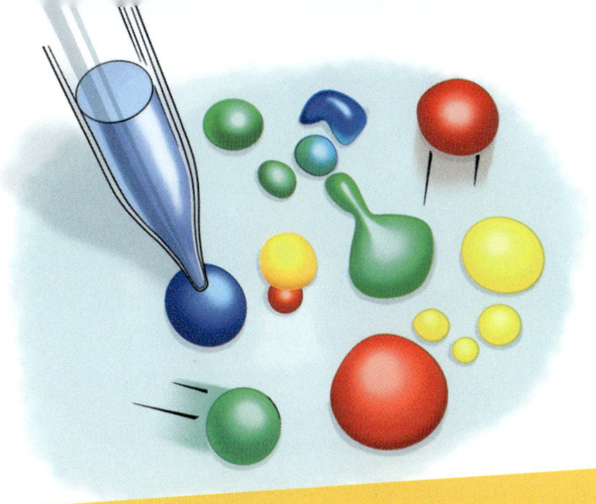

Dancing Droplets

CAN YOU MAKE FOOD COLORING AND WATER MOVE AROUND ON THEIR OWN—AS IF THEY WERE DANCING?

PROJECT TIME

45 to 60 minutes

KEY CONCEPTS

Physics
Mixture
Materials science
Water droplets

Have you ever watched raindrops on your window as they move and run down the glass? It is fascinating to observe how some of them sit there by themselves whereas others combine to build a larger drop. Have you ever wondered what makes them move and behave in different ways? Controlling and influencing liquids spreading on surfaces, known as "wetting," is actually important for many things. For some purposes, you might want a surface to repel a liquid; just think of your water-repellent jacket as one example. Other times you need the liquid to spread evenly across a surface, such as paint on a wall. But can you predict what a drop of water will do on a specific surface? Will it spread or form a droplet? It is even more complex if the liquid is not only water but a mixture of different components. Sometimes the outcome can be very unexpected, and even fascinating to watch. Try this activity and see for yourself!

BACKGROUND

What determines how a liquid behaves on a surface? The answer seems easy: it depends on the liquid and surface. Different surfaces have different molecular structures, which result in different affinities for water or other liquids. Materials that attract water, such as clean glass, are called hydrophilic, which means water-loving, and water spreads easily on them. Other surfaces, such as Teflon or plant leaves, exhibit the opposite behavior and are water-fearing or hydrophobic, meaning water will form a round droplet on their surfaces. The shapes of a droplet on a specific surface can actually tell you a lot about the characteristics of the material it is on.

In liquids, surface tension is the property responsible for the shape of liquid droplets. This is how it works: Liquid molecules inside a droplet want to cling to their surrounding neighbors. This makes the molecules inside the liquid stick together. On the surface, however, there are fewer liquid molecules around, which leads to the formation of stronger bonding between the surface molecules. As they get pulled mostly inward by the molecules below, all the surface molecules help "hold together" to form a droplet. You have probably seen examples of surface tension in action without realizing it: Tiny droplets of water on the window or plant leaves after it has rained or bugs called water striders walking on water are just a few of them. In this activity, you will explore another example of surface tension in action, and you will discover that other factors influence droplet behavior too. Be prepared for some colorful droplet action!

MATERIALS

- Adult helper
- Water
- Paper towel
- White paper
- Microscope glass slide and pliers (optional)
- Heat-tolerant, flat glass surface, such as a Pyrex glass tray or baking dish (in good condition with no cracks)
- Liquid food coloring (that contains propylene glycol as the main ingredient)
- Medicine dropper
- Oven mitts
- Gas stove or a camp stove or Sterno cooking fuel with matches or a lighter (Be sure to use your heat source in an appropriate area with

appropriate precautions; always follow all manufacturer instructions.)

- Yogurt container lid or wax paper
- Safety glasses (to protect experimenters' and observers' eyes)
- Paper towels
- One sheet of white paper
- Permanent marker (optional)

PREPARATION

- Place paper towels or other protective layers that can get dirty on your work area.
- Take the medicine dropper and put a line of five water droplets on a plastic yogurt lid or wax paper. They should all be approximately the same size.
- Choose one color of food coloring (blue, green, red, or yellow) and place one drop of concentrated food coloring from the bottle next to the line of water drops.
- With the medicine dropper, take a small amount of the concentrated food coloring from the drop next to your line and put it into the first water droplet of the line. Mix it by sucking the droplet in and out the dropper.
- Now suck up a little amount of this first droplet with the medicine dropper and mix it into the next fresh water droplet in the line.
- Continue doing this for all five water droplets. At the end, you should have a line with six droplets of food coloring in different concentrations.
- Rinse your medicine dropper with fresh water.

- Repeat the whole mixing procedure with a new line of five water droplets and another selection of food coloring (blue, green, red, or yellow). Make sure to rinse your medicine dropper with fresh water in between colors. Set the yogurt lid aside for now; you will need the food-coloring droplets later in the activity.

- Have your adult helper light the camping or gas stove. If you do not have either one available, light the Sterno with a match or lighter and use this flame. Use extreme caution when handling open flames. Check your environment for safety and tie up loose hanging hair.

- Have everyone put on eye protection.

- Have your adult helper hold the heat-resistant glass dish with the oven mitts. Pass part of the bottom flat area (an area of about 1 by 2 inches [2.5 by 5 cm] is enough) slowly through the gas or camping stove flame for about 30 to 45 seconds. The flame should touch the glass. If you use Sterno, pass the flat glass surface through the flame for about one to two minutes. Make sure the glass touches the flame. Remember where the spot on your dish is that touches the flame—you will use this area for your activity. (Caution: Be careful when passing the glass through the flame. Small microscopic cracks inside the glass can make it shatter when exposed to heat. Make sure to only use a heat-resistant or heat-tolerant glass dish, and all bystanders should be outfitted with eye protection, such as safety glasses.)

- Alternatively, if you have microscope slides, hold them with pliers and pass them in the flame for 30 seconds (gas or camping stove) or one minute (Sterno).

- Use caution, the glass will be very hot. Carefully place the glass dish or microscope slide on your work area. (The side that you passed through the flame should face upward.) Let the glass cool for approximately 10 minutes.

PROCEDURE

- Once your glass has cooled a bit, place a white sheet of paper on your work area. Next, place your still-warm glass dish or microscope slide on the paper. Remember that the area that faced the flame should face upward.

- Using your rinsed medicine dropper, put a small drop of pure water onto the glass surface. *What happens to the drop? Does it spread on the surface, or does it keep its drop shape?*

- With the medicine dropper, now take a small droplet of concentrated food coloring from your yogurt lid or wax paper (pick any color) and carefully place it on the glass. *How does this droplet behave? Does it look different than the pure water droplet? Do you notice any other difference between the two drops?*

- Now put a droplet of the same concentrated food coloring next to the other concentrated food-coloring droplet on the glass. They should not touch each other, but should be in close proximity (about 0.1 to 0.2 inch [2.5 to 5 mm] apart). *How do the two droplets behave? Are they moving or are they still? If they are moving, in which direction do they move and what happens to them?* (**Tip:** If you do not see anything happening, you probably haven't heated the glass surface enough. Wipe off the glass and try to pass the glass surface through the flame again, making sure to concentrate the flame to one area and give it enough time there. It works best with microscope slides, but flat surfaces of a heat-resistant glass dish are also fine to use. You can also try to put the droplets on a different spot of the heated glass surface—some areas might work better than others.)

- After rinsing your medicine dropper, take another droplet of concentrated food coloring (pick any color) and place it on the warm glass surface.

- Make sure to rinse your medicine dropper again. Then, from your line of food-coloring droplets on the yogurt lid or wax paper, pick the least concentrated one of a different color and put a small droplet next to the concentrated food-coloring droplet on the glass so neither drop touches. *Do these two droplets show the same behavior as the two concentrated food-coloring droplets? Do you see any differences? How do they behave differently?*

- Now pick several colors and concentrations of food coloring from your yogurt lid or wax paper and place many different small droplets randomly on your warm glass surface. Make sure to rinse your medicine dropper before you switch colors or concentrations. Try to also make different droplet sizes. *What happens to the different droplets? Does the color or concentration change their behavior? Can you make out any behavioral patterns?*

SCIENCE FAIR IDEA

Can you think of other ways to influence the behavior of your droplets on the glass or any other surface? They will not dance on every surface, but you can try to play around with droplet formation and spreading. Think of surfaces such as the fabric of your rain jacket, plant leaves, plastic, or a wooden table. *Will treating the surfaces with soap, olive oil, or alcohol change how droplets form? What happens if you change the surface's roughness?* (Food coloring and oil-based substances can stain, so ask permission before trying these liquids on different surfaces.)

OBSERVATIONS AND RESULTS

You should have seen all kinds of droplet behaviors on the glass surface. The droplet of pure water probably spread out pretty fast. Treating the glass dish or microscope slide with the flame made the surface very unfavorable for droplet formation and more susceptible for water spreading. The concentrated food coloring, however, contains not only water but also a chemical called propylene glycol. The mixture of both components allows the formation of a droplet that is more sphere-shaped, which you should have noticed when putting the concentrated food coloring on your glass surface. The water on the droplet surface evaporates faster than propylene glycol, and at the same time, has a higher surface tension. These differences result in a symmetric internal water flow inside the droplet that stabilizes it and stops spreading.

You might have seen that the food-coloring droplet was floating or moving on the glass a little bit. When putting a second food-coloring droplet next to the first one, they glide over the glass and move toward each other. This attractive motion is vapor-mediated, meaning that one droplet influences the water evaporation of the second droplet, leading to an imbalance of the stabilizing internal water flow. The resulting asymmetry starts the movement of the droplets toward each other. Once the two droplets meet, their water/propylene glycol mixture determines their behavior. If the mixture is the same, the two droplets join to become a bigger droplet, as you might have seen. If you choose a very different food-coloring concentration for your second droplet, you can observe one drop chasing the other. The reason for this chasing behavior is the difference in surface tension of the two droplets. Lower-surface-tension droplets (droplets with a higher concentration of food coloring) will chase the drop with higher surface tension (droplets that have lower food-coloring concentration). This makes for some fun observations: Did you see your droplets dancing once you put many different droplets with lots of different colors and food coloring concentrations on your warm glass surface?

CLEANUP

Rinse all your glassware or microscope slides with water to get rid of the food coloring. Wash all the other materials you used (yogurt lid, medicine dropper) and recycle the yogurt lid if you are not planning to reuse it. Clean your work area with water and soap and do not forget to wash your hands.

The Magnifying Effect of a Water Drop

CAN A DROPLET OF WATER CHANGE WHAT YOU CAN SPY WITH YOUR LITTLE EYE? LEARN ABOUT OPTICS AND LENSES IN THIS LARGER-THAN-LIFE SCIENCE ACTIVITY!

PROJECT TIME

15 to 20 minutes

KEY CONCEPTS

Lens
Magnifying glass
Optics
Light

Have you ever studied an everyday object through a magnifying glass—and been amazed at what you could see? Or have you ever noticed, for example in a swimming pool, that an object that is sticking out of the water looks different just above and just below the surface? In this activity, you will learn a little bit more about both of these observations. Get ready to bend light, magnify letters and have fun with water drops—all while getting a glimpse into how lenses work! Lenses are the key components in eyeglasses, contact lenses, binoculars and telescopes—just to name a few devices. With this activity, a homemade magnifying glass is only a drop away!

BACKGROUND

You see an object because light rays reflected from the object shine into your eye, creating an image on the retina inside your eye. Signals to your brain allow it to re-create the picture of the object.

A light ray bouncing off an object usually travels in a straight line to your eye. Things change when a transparent material, such as glass or water, gets in the way. When a light ray traveling through air enters such a material, it changes direction, creating a sort of kink. Another kink is introduced when the ray leaves the material. Therefore, the final image of the object in your eye might be different due to the changes in direction of the light on its way to your eye. Your brain is unaware of these kinks and expects an image created by rays that traveled in a straight line. As a result, it might reconstruct a picture that is different from the initial object. Your eyes and brain might have been fooled!

Lenses use these kinks to make objects look bigger or smaller, closer or farther away. A convex lens bends light rays inward, which results in the object being perceived as larger or closer. A concave lens bends rays outward; you get the perception that objects are smaller or farther away. There is no overall bending of light for a flat lens. You perceive the object as it is.

Now that you know a little about light and lenses, are you ready to let your eyes and brain be fooled?

MATERIALS

- A newspaper page
- Two metric rulers, preferably with dark markings
- One transparency film or clear sheet protector
- Drinking glass with water
- Medicine dropper (optional)
- Mobile device with a camera (optional)
- Small transparent plastic cup or tiny tasting cup with a flat bottom (optional)

PREPARATION

- Find a waterproof work area.

- Select an article in the newspaper with a small font. You can use your ruler to measure the height of the letters; they should be a few millimeters high.

PROCEDURE

- Place the transparency film on top of a newspaper page.
- Create a drop of water near the middle of the transparency film. Use a water dropper or your finger to let two or three drops fall on the film and merge into one bigger drop. Examine your water drop. *Is the top of the drop surface flat, curved inward, or curved outward?*
- Shift your transparency film so the water drop lays on top of the small print letters.
- Close or cover one eye and look from above with the other eye at the letters under the drop. Compare them with the letters next to but not covered by the drop. *Do they look the same? Does one appear bigger or smaller than the other?*
- Using two hands, carefully lift and hold the transparency film about 1/2 inch (1.3 cm) above the newspaper, leaving the newspaper on your work surface. You might need help lifting the transparency film if you like to cover one eye with a hand.
- Close or cover one eye and look carefully from above through the water drop at the letters on the newspaper. *Do the letters appear different than when the transparency film rested on the newspaper? What happens when you move the transparency film farther up?*
- Move the transparency film up and down a couple of times looking from above through the water drop with one eye. *How does your perception of the letters change as you move the transparency film farther up or back down? Why do you think this happens?*

- To measure the magnification factor of your water drop, put a ruler under your transparency film on your work surface and another ruler next to the drop on top of the transparency film, but be sure to prevent the ruler from touching the drop.
- Lift the transparency film with the top ruler and water drop about 0.5 inch (1.3 cm) up and do your best to measure the length of a millimeter indication of the bottom ruler, as seen through the water drop. (You might need help lifting the transparency film together with the ruler and the water drop.) *How many millimeters does 1 millimeter indication measure?* This number tells you by what factor objects appear bigger when seen through your water drop. *Are you surprised about the magnification factor you obtained?*
- Measure the magnification factor of your water drop when you lift the transparency film higher up. *Does the magnification factor change when you lift the transparency higher? Could you find ways to make the magnification factor very big?*
- Repeat the activity, this time using a larger water drop. *What happens to the curvature of the top surface of the water drop when you increase the size of the drop? Is it more, less, or similarly curved? Do bigger water drops yield a different magnification factor?*

EXTRA

What do you think would be the optimum water drop size and its height above the newspaper to increase the readability of your chosen newspaper line? Would you choose the same conditions if you were investigating the details of an insect?

EXTRA

Go around the house or your yard looking at objects through your new magnification glass. *What kind of surprising details can you find?*

EXTRA

You just used water to create a magnifying glass, making objects appear larger. *What do you think will happen if you look through a bigger layer of water held in a cup?* To test this, find a small or tiny transparent plastic cup with a flat bottom. To verify that the cup itself does not act as a lens, place the empty cup over a straight line in your newspaper, look through the cup, and observe. *Does the line appear straight? Does it appear to have the same thickness if you lift the cup?* If not, find another cup, because the bottom of this cup already acts as a lens. *Why is it important that the cup used to test if a layer of water in a cup acts as a lens does not act as a lens already?* Once you have a cup that does not act like a lens, fill it with a layer of water (0.5 inch [1.3 cm] deep) and look from above through the water to the letters on your newspaper. *How do the letters appear? Does their appearance change when you move the cup up and down?* You might want to shift to a font with larger letters. *Can you calculate the magnification factor of this lens?* Note that a magnification factor smaller than 1 indicates the object appears smaller than it is. As an example, a magnification factor of one-half indicates that the object appears to be half its size.

EXTRA

Optical instruments often use a combination of several lenses. If you made both the water-drop lens and the cup-with-water lens, observe what happens if you combine both. You can put the transparency on top of your cup and look from above or ask a helper to hold your transparency film with the water drop as you hold the cup above it. *What do you think will happen? Can you measure the magnification factors for both lenses individually (at the exact distance you are holding them when you combine both) and when combined?* If you do this for a couple different distances, you might be able to develop a formula.

SCIENCE FAIR IDEA

Would other liquids also create magnification? Would one liquid work better than another? Think of oil or vinegar or soy sauce. *Which ones do you think might work, and why?*

OBSERVATIONS AND RESULTS

Did you see how objects appear larger when looked at through a water drop? The surface of a water drop curves outward to make a dome. This outward, or convex, curvature bends light rays inward. The result is an enlarged image on the retina of your eye. The object appears bigger than it is.

The surface of a smaller drop is even more curved, creating a bigger change in direction of the light ray. The result is a larger magnification. Changing the position of the water drop with respect to the letters and your eye will also affect the magnification factor. Due to something called the capillary effect, however, a layer of water in a cup shows a surface that is slightly bent inward. It will act as a concave lens that bends the light rays outward. As a result, letters seen through the layer of water in a cup appear smaller than they are. When you combine several lenses, the magnification factor of the set of lenses is the product of the magnification factors of the individual lenses.

Any clear liquid will work as a lens. As long as the bottom and top surfaces of the layer or drop are not parallel, the lens will change the appearance of the object. Depending on the liquid, the magnification factor of similar drops made up of different liquids will vary.

CLEANUP

Use a towel to wipe up any water on your workspace and return all materials to where you got them.

Salty Science
Floating Eggs in Water

SALT IS FOR MORE THAN JUST SEASONING YOUR FOOD! DID YOU KNOW IT CAN ALSO HELP YOUR FOOD—AND OTHER OBJECTS—FLOAT IN WATER? TRY THIS ACTIVITY TO DISCOVER HOW SALT WATER AND FRESH WATER AFFECT BUOYANCY.

PROJECT TIME

30 to 45 minutes

KEY CONCEPTS

Density
Mass
Volume
Concentration
Buoyancy
Water

Have you ever wondered why some objects float in water and others sink? It has to do with the density of the objects compared with the density of the water surrounding them. If an object is less dense than the water around it, it will float. Because salt water is denser than fresh water, some things float more easily in the ocean—or extremely salty bodies of the water, such as the Dead Sea. You can make your own dense water by adding salt to tap water. In fact, if you add enough salt, you can make the water so dense that an egg will actually float in it! Explore how this works in this science activity.

BACKGROUND

If you put an egg in a cup of tap water, it will sink to the bottom. Why is this? Because the density of the egg is higher than the density of tap water, so it sinks. Density is the mass of a material per unit volume. For example, the density of fresh water under standard conditions is approximately 1 gram per cubic centimeter.

But, if you add enough salt to the water, the egg will actually float back up to the surface! Adding salt to the water increases the density of the solution because the salt increases the mass without changing the volume very much.

When enough salt is added to the water, the saltwater solution's density becomes higher than the egg's, so the egg will then float! The ability of something, like the egg, to float in water or some other liquid is known as buoyancy. But just how much salt is needed to make an egg float? In this science activity, you'll figure that out by making solutions with varying concentrations of salt in them.

MATERIALS

- One egg
- Water
- Measuring cup
- Large container, such as a large bowl or cooking pot (It must be able to hold at least 3 cups.)
- 1/2 cup of table salt
- Five cups that hold at least 16 ounces (473 ml) each
- Permanent marker (if you are using plastic cups) or masking tape and a pen (to label reusable cups)
- Three spoons for mixing salty solutions
- Soup spoon for egg transfers

PREPARATION

- Take the egg out of the refrigerator and allow it to warm to room temperature. Be sure to always wash your hands after handling uncooked eggs because they may carry salmonella.
- Pour 1 1/2 cups (355 ml) of water into your large container.
- Add 1/2 cup of salt to the large container and stir to dissolve some of the salt (it will not all dissolve yet).

- Add 1 cup (237 ml) of water to the large container (making 2 1/2 cups [0.6 L] total) and stir to dissolve the remaining salt. The salt should be completely dissolved before you go on to the next step. It may take several (up to 10) minutes of stirring, so you may need to be patient. *Why is it important to start out with a solution that has such a high concentration of salt?*
- Arrange the five cups on a surface, going in a line from left to right. Label the cups 1 to 5. If you are using plastic cups, you can use a permanent marker to label them. If you are using reusable cups, you can use masking tape and a pen to label them.

PROCEDURE

- Add 3/4 cup (177 ml) of the salty solution you prepared to cup 1.
- Add 3/4 cup (177 ml) of plain tap water to cups 2 through 5. (Cup 5 will be plain tap water.)
- Add 3/4 cup (177 ml) of the salty solution you prepared to cup 2 and mix it. *What is the salt concentration in cup 2 compared with cup 1?*
- Add 3/4 cup (177 ml) of the salt solution from cup 2 to cup 3 and mix it. *What is the salt concentration in cup 3 compared with cups 1 and 2?*
- Add 3/4 cup (177 ml) of the salt solution from cup 3 to cup 4 and mix it. *What is the salt concentration in cup 4 compared with the other cups?*
- Use a soup spoon to place an egg in cup 5. *Does the egg float?*
- Use the spoon to take the egg out and place it in cup 4. *Does the egg float?*
- Repeat this process with cups 3, 2, and then 1. *In which cup does the egg first float? If the egg floated in more than one cup, did you notice any difference in how it floated? What does this tell you about the density of the egg?*

EXTRA

In this science activity, you figured out, within a factor of two, how much salt it takes to float an egg. You could narrow down the range further by testing additional saltwater solutions to try and determine the egg's density. To do this, start your solution with the salt concentration in which the egg first floated and make a new dilution series, as you did before. *Now in which cup does the egg first float? What does this tell you about the density of the egg?*

EXTRA

Repeat this activity using several more eggs, possibly both hard-boiled and uncooked eggs. *Do you get the same results with other eggs or is there some variation between different eggs?* For testing hard-boiled versus raw eggs, you should test the same egg, first raw and then after hard-boiling it, to investigate any differences.

EXTRA

Find out how much salt there is in seawater. *From the results of your activity, do you think an egg would float or sink in seawater?*

OBSERVATIONS AND RESULTS

Did the egg float in cups 1 and 2, but not in cups 3, 4, or 5?

You likely saw that the egg floated best in cup 1, floated a little less in cup 2 (but part of it was above the surface), and did not float in the other cups. Cup 1 had the undiluted salty solution that you originally prepared, which was 1/2 cup of salt in 2 1/2 cups (0.6 L) water total. The concentrations of the salt solutions in cups 2 to 4 were halved as you increased in cup number; so, the concentration of the salt in cup 2 was half that of cup 1, and the concentration of the salt in cup 3 was half again of cup 2. (Cup 5 had plain tap water.)

The egg should have sunk in cups 3, 4, and 5 because the density of the egg was higher than the density of the solutions (or plain tap water) in those cups. Cups 1 and 2 had more salt in them than the other cups (with cup 1 having the most salt), which means these solutions were denser. The egg should have floated (with part of it above the water surface) in these two cups because the solutions were denser than the egg. The actual density of the egg is in between the density of the solution in cup 3 and that in cup 2.

CLEANUP

Pour out the water, and clean the containers. Throw out the egg if it's been out of the refrigerator for longer than two hours, otherwise you can cook and eat it. Return all items to where you found them.

Floating with a Swim Bladder

FLOAT LIKE A FISH—WITH A LITTLE BIT OF PHYSICS!

PROJECT TIME

15 to 20 minutes

KEY CONCEPTS

Physics
Water
Buoyancy
Archimedes' principle

Have you ever thought about why most fish never sink to the bottom of the ocean or float to the water's surface? How is it that they can stay so perfectly buoyant underwater? You might be surprised to hear most bony fish have a special organ to help them with that: a swim bladder. This is a thin-walled sac located inside the body of a fish that is usually filled with gas. Besides helping fish stay buoyant, it can also function as a sound producer and receptor or as an accessory respiratory organ. Wonder how a swim bladder works? Try this activity to find out!

BACKGROUND

You probably know objects can sink or float when placed into water. In both cases, the water moves out of the way, or is displaced, to make space for the object. An ancient Greek scientist named Archimedes discovered this principle of buoyancy. In physics, buoyancy is an "upward force that pushes on an object that is immersed in a liquid." If you have ever tried to push a beach ball underwater, you have felt this buoyancy force in action. The ball seems to push back up although you are pushing it down! Archimedes also discovered the strength of the upward acting force exerted by the liquid is equal to the amount of water displaced by an object. At the same time, the volume of displaced water is always the same as the volume of a submerged object. This is why the water-displacement method is a great way to determine the volume of an irregularly shaped object.

Whether an object sinks or floats is dependent on which force is larger: the force of gravity pulling down on the object (its weight) or the buoyancy force pushing up on it. If the buoyancy force is larger than the object's weight, it will float on the water's surface. If the object's weight exceeds the buoyancy force, however, it will sink to the bottom. When the buoyancy force is exactly the same as the object's weight, the object has neutral buoyancy and remains at its level. This means the object's density becomes important as well. Dense objects have a high mass and only displace a low volume of water; they have a large mass-to-volume ratio. If an object has a small mass but displaces a lot of water, it is less dense and therefore has a low mass-to-volume ratio. Objects denser than water will sink whereas those less dense than water will float.

What does all this have to do with a swim bladder? For a fish to be buoyant, or float, it must displace an equal or greater amount of water than its own body mass. The trick is the swim bladder, which is basically like an air-inflated balloon that can expand and contract depending on how much gas is inside. When the swim bladder expands it will increase in volume and therefore displace more water. This increases the fish's buoyancy, and it will float upward. When the swim bladder deflates, the fish's buoyancy decreases and it will sink as it displaces less water. Divers use the same concept for their buoyancy-control devices.

MATERIALS

- Small glass bottle
- Balloon that fits inside the bottle
- Plastic tubing (at least 12 inches [30.5 cm] long)

- Big tub or container that is waterproof and preferably transparent (It should have enough room so the glass bottle can float in it freely.)
- Strong, waterproof tape
- Water
- Towels

PREPARATION

- Find a workspace that can withstand water spills.
- Push one end of the tube through the opening of the balloon, and tape the balloon and the tube together, creating a seal. After that you should be able to inflate the balloon by blowing through the tube. Make sure the connection between the balloon and tube is airtight.
- Put the balloon inside the glass bottle and tape the tube to the bottle opening.
- Fill the tub with water.

PROCEDURE

- Place the bottle with the balloon inside into the tub of water. Keep the other end of the tube outside of the tub. *What happens when you place the bottle into the water? Does it float or sink? Why?*
- Carefully blow through the tube so the balloon inside the bottle inflates a little bit. *Does inflating the balloon a little bit change the floating behavior of the bottle?*
- Inflate the balloon even more until it almost fills the whole bottle. *What happens as you inflate the balloon more and more? Can you explain your observations?*
- Next, let the balloon deflate. *How does that change the position of the bottle in the water?*

- Play around with inflating and deflating the balloon inside the bottle. *Can you find a correlation between the balloon size inside the bottle and the latter's floating behavior?*

SCIENCE FAIR IDEA

Try simulating different sizes of fish by using different types of bottles or balloons. *Do all of them give you the same results? What if you put a tiny balloon in a very big bottle or vice versa?*

EXTRA

Does the activity also work with a plastic bottle instead of a glass one? Why or why not? Test it to find out!

OBSERVATIONS AND RESULTS

Can you see how what you built resembles a fish? The glass bottle mimics the fish's body, and the balloon represents the swim bladder inside its body. When you put the bottle into the tub, it should have filled with water and sunk to the bottom. This is because the weight of the glass bottle filled with water is larger than the buoyancy force pushing it up. This changes when you inflate the balloon; when it expands it pushes the water out of the bottle. You now have an air-filled balloon inside the bottle. Because air is much lighter than water, the bottle's weight lessens as the balloon fills with more air. The bottle eventually becomes lighter and thereby exerts a weaker downward push than the buoyancy force pushing the bottle up. At this point, the bottle will start floating upward toward the water's surface.

When you deflate the balloon again the reverse happens—the bottle becomes heavier due to the water flowing back into the bottle, and it starts to sink. You might have noticed the activity doesn't work as well with a plastic bottle. This is because plastic itself is less dense than water and tends to float whereas glass is denser than water and therefore sinks.

CLEANUP

Detach the tubing from the bottle and the balloon and dry all water spills and materials with a towel or paper towels.

Buoyant Science
How Metal "Boats" Float

LEARN HOW BIG SHIPS STAY AFLOAT WITH A LITTLE ALUMINUM FOIL AND SOME WATER!

PROJECT TIME

30 to 45 minutes

KEY CONCEPTS

Archimedes' principle
Hydrodynamics
Fluid dynamics
Physics
Water

Have you ever wondered why when you drop a steel nail into water it sinks like a stone, but when a well-built steel ship is in the ocean, it floats, even though it weighs much more than a tiny nail? The answer has to do with the fact that when an object is placed in water, water is pushed out of the way. You may have noticed this when you take a bath in a bathtub. Known as Archimedes' principle, as water is pushed away by an object, the water exerts a force back on the object that is equal to the object's weight. This is what helps make an object float.

BACKGROUND

More than 2,200 years ago, a scientist named Archimedes sat down in his bath and figured out that when an object is placed in water, water moves out of the way—it gets displaced. If the object is floating, the amount of water that gets displaced weighs the same as the object. There is a force, called a buoyant force, which pushes on an object when it displaces water. The strength of this upward acting force exerted by water is equal to the weight of the water that is displaced. So, if an object displaces just a little bit of water, the weight of that small amount of water is small, and so the buoyant force is small too. If, on the other hand, the object displaces a lot of water, then there will be a large buoyant force pushing upward. Whether an object sinks or floats depends on its density and the amount of water it displaces to create a strong enough buoyant force. How dense can an aluminum "boat" be before it sinks?

MATERIALS

- Cloth towel (or paper towels)
- Large, clear bowl or container
- Water
- Aluminum foil
- Metric ruler
- Pen
- Scissors
- Permanent marker
- Hammer or mallet

PREPARATION

- Spread out the towel or paper towels on a hard work surface. Fill the bowl or container about two-thirds full of tap water and set it on the towel(s).
- Measure out a square of aluminum foil that is about 25 centimeters square. Cut out the square. This will become the metal "boat" you test.
- Mark the four corners of the aluminum foil square with permanent marker.

- **Be careful:** Later in the experiment you will use the hammer. Be sure to pound on a surface that can safely withstand the force and is resistant to damage.

PROCEDURE

- Pull the corners of the aluminum foil square together and crumple the square into a loose ball that is approximately 6 centimeters in diameter. Rumple the aluminum such that the marked corners stay together and are visible in one spot.
- Set the ball gently in the bowl of water, placing it so that the marked corners are at the top of the ball, as this will help prevent the ball from filling up with water. Immediately observe the ball. *Does it sink or float?*
- Get down low so that you are at eye level with the aluminum foil ball and quickly observe how much of the ball is below the surface of the water. *Is about 10 percent, 25 percent, 33 percent, 50 percent, 67 percent, 75 percent, 90 percent, or 100 percent of the ball underwater?*
- Remove the ball from the bowl, shake out any water and dry it on the towel.
- Now crumple the ball a little more tightly, into one that is approximately 5 centimeters in diameter. If you crumple it too much, just carefully pull apart some of the aluminum foil to get the desired size.
- Again, set the ball gently in the water, placing the marked corners at the top. *Does it sink or float? What percentage of the ball is below the top of the water?* Remove it, shake out any water and dry it.

- Continue to crumple the ball to be smaller and tighter, and test whether it floats or not (as you have been doing) as it gets smaller. Keep testing smaller diameters until the ball completely sinks. Try testing these diameters (or similar): 4 centimeters, 3 centimeters, 2.5 centimeters, 2.25 centimeters, 2 centimeters, 1.75 centimeters, 1.5 centimeters. If it is too hard to squeeze the ball smaller by hand strength alone, then carefully use the hammer or mallet to gently pound the foil into a smaller ball (or as close to a ball-shape as you can make it). *For each diameter you test, what percentage of the ball is submerged?*
- *At which diameter did the ball sink to the bottom? Do you think that the ball that sank had the lowest or highest density? At which diameter did the ball have a density that was approximately equal to that of water? (When was the ball almost completely submerged or fully submerged but not quite sinking to the bottom?)*

EXTRA

Cut out at least two additional aluminum foil sheets that are 25 centimeters square and repeat this activity. *Do you get the same results with all the squares you test, or is there a lot of variation?*

EXTRA

You can do this activity again, but this time weigh the aluminum sheet on a scale and calculate its mass in grams. Calculate the volume of the spheres for each diameter, using the fact that the volume of a sphere is equal to 4/3 times pi (3.14) times the radius cubed. Using the mass and the volumes, compute the average density of the aluminum sheet for each diameter by dividing mass by volume. *At what density did the aluminum ball sink? At what density was the aluminum ball approximately equal to that of water? For each diameter of the sphere, what is the mass of the water that was displaced?* For more accurate results, continue testing additional 25-centimeter aluminum squares.

OBSERVATIONS AND RESULTS

Did more and more of the ball end up below the top of the water as the ball's diameter decreased? Was about half of the ball below the water when the ball had a diameter of about 2.5 centimeters, and did the entire ball sink when its diameter was about 1.5 centimeters or smaller?

If an object is floating in water, the amount of water that gets displaced weighs the same as the object. Consequently, while it was floating, the ball should have displaced the same amount of water as it decreased in diameter, and so the buoyant force should have remained the same. However, the density of the ball was changing—it increased as the ball's diameter decreased.

Density is the mass per unit volume—it describes how much "stuff" is packed into a volume of space. When the aluminum ball had a diameter of 5 centimeters, the ball should have floated well because it had a density lower than that of water due to the air inside of the ball, just like steel ships that can float because their density has been lowered by encasing air inside the hull. And as long as the ship displaces enough water to create a strong buoyant force, it can stay afloat—even if it is loaded with cargo. As the diameter decreased and density increased, the ball should have sunk more and more. When its diameter was about 1.75 centimeters or 1.5 centimeters, you may have seen it become 90 percent submerged. This is when the ball had a density approximately equal to that of water. With a diameter of about 1.5 centimeters or smaller, the ball should have sunk, indicating that its density was greater than that of water, thereby overcoming the buoyant force.

CLEANUP

Pour the water down a drain and recycle the aluminum foil.

THE SCIENTIFIC METHOD

The scientific method helps scientists—and students—gather facts to prove whether an idea is true. Using this method, scientists come up with ideas and then test those ideas by observing facts and drawing conclusions. You can use the scientific method to develop and test your own ideas!

Question: What do you want to learn? What problem needs to be solved? Be as specific as possible.
Research: Learn more about your topic and refine your question.
Hypothesis: Form an educated guess about what you think will answer your question. This allows you to make a prediction you can test.
Experiment: Create a test to learn if your hypothesis is correct. Limit the number of variables, or elements of the experiment that could change.
Analysis: Record your observations about the progress and results of your experiment. Then analyze your data to understand what it means.
Conclusion: Review all your data. Did the results of the experiment match the prediction? If so, your hypothesis was correct. If not, your hypothesis may need to be changed.

GLOSSARY

accelerate: To increase in speed.

compost: A mixture of decaying plants such as leaves, cut grass, and food scraps from plants. It's added to soil to make it better.

concave: Having a shape like the inside of a bowl.

convex: Having a shape like the outside of a bowl.

diameter: The distance from one side of a round object to another through its center.

immerse: To put something into matter, such as a liquid, that completely surrounds it.

integral: Essential or important.

phenomenon: A fact or an event that is observed.

physics: The study of matter, energy, force, and motion, and the relationship among them.

porous: Having small holes that allow air or water through.

respiratory: Having to do with the movement of air in and out of the lungs.

retina: The part of the eye that converts an image into a signal that's sent to the brain.

susceptible: Being open to some influence.

water-repellent: The ability to drive away water or other liquids.

ADDITIONAL RESOURCES

Books

Andrus, Aubre and Rachel Rothman. *Good Housekeeping Amazing Science: 83 Hands-On S.T.E.A.M. Experiments for Curious Kids!* New York, NY: Hearst Home Kids, 2021.

Castaldo, Nancy F. *When the World Runs Dry: Earth's Water in Crisis.* Chapel Hill, NC: Algonquin Young Readers, 2022.

Heinecke, Liz Lee. *Physics for Kids: Science Experiments and Activities Inspired by Awesome Physicists, Past and Present.* Beverly, MA: Quarto Publishing Group USA Inc., 2022.

Websites

Discovery Education
sciencefaircentral.com

Exploratorium
www.exploratorium.edu/explore/activities

Science Buddies
www.sciencebuddies.org/science-fair-projects/project-ideas/list

Science Fun for Everyone!
www.sciencefun.org/?s=science+fair

Videos

Science Trek: Water
www.pbs.org/video/science-trek-water/

Spit: Everything You Never Wanted to Know
www.pbslearningmedia.org/resource/spit-its-okay-to-be-smart/spit-its-okay-to-be-smart/

Publisher's note to educators and parents: Our editors have carefully reviewed these websites to ensure that they are suitable for students. Many websites change frequently, however, and we cannot guarantee that a site's future contents will continue to meet our high standards of quality and educational value. Be advised that students should be closely supervised whenever they access the internet.

INDEX